Also available from ASQ Quality Press

ISO 9000 Quality Management System Design: Optimal Rules for Documentation, Implementation, and System Effectiveness
Jay J. Schlickman

Food Processing Industry Quality System Guidelines
ASQ Food, Drug and Cosmetic Division

Product Development Planning for Health Care Products Regulated by the FDA
Elaine Whitmore

Practical Product Assurance Management
John Bieda

Failure Mode and Effect Analysis: FMEA from Theory to Execution
D.H. Stamatis

To request a complimentary catalog of ASQ publications, call 800-248-1946.

Contents

The Product Recall Planning Guide

Second Edition

American Society for Quality
Product Safety and Liability Prevention Interest Group

ASQ Quality Press
Milwaukee, Wisconsin

The Product Recall Planning Guide—2nd ed.
[Prepared by] American Society for Quality, Product Safety and Liability Prevention Interest Group.

Library of Congress Cataloging-in-Publication Data

The product recall planning guide / [prepared by] American Society for Quality, Product Safety and Liability Prevention Interest Group. — 2nd ed.
p. cm.
Includes bibliographic references and index.
ISBN 0-87389-457-X (alk. paper)
1. Product recall—United States—Management. I. American Society for Quality. Product Safety and Liability Prevention Interest Group.
HF5415.9.P76 1999
658.5′6—dc21 99-17454
CIP

10 9 8 7 6 5 4 3 2 1

ISBN 0-87389-457-X

Acquisitions Editor: Ken Zielske

Project Editor: Annemieke Koudstaal

Production Coordinator: Shawn Dohogne

ASQ Mission: The American Society for Quality advances individual and organizational performance excellence worldwide by providing opportunities for learning, quality improvement, and knowledge exchange.

Attention: Bookstores, Wholesalers, Schools and Corporations:
ASQ Quality Press books, videotapes, audiotapes, and software are available at quantity discounts with bulk purchases for business, educational, or instructional use. For information, please contact ASQ Quality Press at 800-248-1946, or write to ASQ Quality Press, P.O. Box 3005, Milwaukee, WI 53201-3005.

To place orders or to request a free copy of the ASQ Quality Press Publications Catalog, including ASQ membership information, call 800-248-1946. Visit our web site at http://www.asq.org.

Printed in the United States of America

Printed on acid-free paper

American Society for Quality

ASQ™

Quality Press
611 East Wisconsin Avenue
Milwaukee, Wisconsin 53202
Call toll free 800-248-1946
http://www.asq.org
http://standardsgroup.asq.org

The Product Recall Planning Guide

Second Edition

Foreword

The recall of a product is an expensive and traumatic incident for the manufacturer and the user. The manufacturer has a responsibility to remove the product from circulation and use to prevent damage and injury. The user may find that he must stop using an essential item because of a serious hazard. How long will his equipment be out of service? It is to everyone's benefit that this be done expeditiously and economically. In some instances government agencies can force the recall of a product. In these cases and others it is in the best interest of the manufacturer to see that the recall be promptly performed. In all cases the manufacturer bears the cost, therefore it is advantageous that the task be performed as economically as possible. The economies are the result of handling the problems expeditiously: by removing or correcting the defects from the field that may cause damage or compensable injury. Major improvements in the economy of recall can result from preplanning. It is the purpose of this guide to assist in preplanning.

The Product Safety and Liability Prevention Interest Group acknowledges with sincerity the contributions of many individuals who have had an interest in liability and product recall and who have contributed to this guide. The many individuals who have made their comments and contributed their time to review and to suggest improvements and changes in the original 1981 edition of this document deserve to know that their efforts have been appreciated.

Despite the changes in technology, many of the earlier suggestions and procedures are still pertinent, but other procedures have grown and changed. The current document is set forth as a guide to assist individuals and organizations in setting up procedures that will assist in their endeavors to keep their product safe and their users protected by providing a way to recall and eliminate from circulation a product that is fundamentally unsatisfactory.

Introduction

Product safety is, without question, one of the most critical aims of a manufacturing organization. Yet, even with our best efforts, there is always the potential for unforeseen problems that may ultimately result in having to recall the product.

An effective plan for product recall planning is a must for any company that has even the remotest possibility of product safety exposure. In today's marketplace, the expectation is to purchase a product safe enough to be used in almost any circumstance, and by even the most incompetent of individuals. The consumer or end user of a product is demanding legislative protection from unsafe products, and is more aggressive in pursuing grievances by instituting legal action against the manufacturer or subcomponents of the product.

Defective materials, failure to properly warn the user of hazards, improper design, and nonconformities of manufacturing specifications are some of the reasons that can cause a product liability situation. All of these must be addressed in a preventive mode if we are to minimize our exposure to unsafe products leaving our facilities, and possibly a subsequent product recall.

Many people in large organizations today simply think procuring product liability insurance will take care of the liability responsibility for unsafe products. This only provides protection against financial losses due to claims and lawsuits and normally does not protect against the high cost of handling a product recall. This is why, in addition to maintaining a comprehensive prevention system for lessening the potential for producing defective products, manufacturers must also be ready with an immediate recall plan to effectively swing into action when the situation dictates that a recall is necessary.

Product liability insurance does little to show evidence that a sound program has been installed to minimize, or hopefully eliminate, unsafe products from entering into the marketplace.

This product recall planning guide is designed not only to assist an organization in planning the recall process, but also to provide methodologies for product liability prevention.

It was with the above thoughts in mind that a program was started by the Product Safety and Liability Prevention Group to replace the original guide that had been published in 1981. Not only was the original edition out of print, but it was possible to misinterpret many areas in the publication. The original effort was begun by Richard M. Jacobs. This issue of *The Product Recall Planning Guide* is the product of a number of members of the Product Safety and Liability Prevention Technical Group. Every effort was made to preserve the useful portions of the original publication.

The primary contributors were

Everett H. Dale, Internal Consultant, Donaldson & Co., 6712 Galway Drive, Minneapolis, MN 55439

Randall Goodden, Vice President, Everbrite Inc., P.O. Box 20020, Greenfield, WI 53220

Elmer (Bud) Gookins Jr., Executive Vice President, General Plug and Manufacturing Co., P.O. Box 26, Grafton, OH 44044

Paul D. Krensky, Paul D. Krensky Associates Inc., 9 Meriam Street, Adams Building, Lexington, MA 02173

Charles F. Seyboldt, Weinstein Associates, 8 Lincoln Street, Brunswick, ME 04011

August B. Mundel, Chair, August B. Mundel, P.E., 34 Sammis Lane, White Plains, NY 10605

Part I

Planning for Possible Recalls

Chapter I

Management Aspects of Recall Planning

General Guideline—Scope

The recall of a product is generally a management decision. An exception can exist when a recall is forced by a government agency. Recalls, in general, are based on a company's total commitment to product safety, moral/ethical, legal/statutory/regulatory requirements, and financial considerations. A recall can be for the purpose of destroying, replacing, or altering the product. Under each of these conditions, the recall decision is made by the company's management.

Recalls are made to protect the customer and the corporation. In the event of a mandatory recall—one dictated by a government agency—the same basis applies. It is done to protect the customer and it also protects the corporation from many additional judicial efforts that could be initiated by customers to obtain redress for damages.

The performance of a recall procedure requires manpower and financial resources. The recall is likely to be more successful and less expensive if a plan and a traceability system exist. It is the purpose of this guide to assist in the establishment of adequate planning and traceability so that a recall, if required, can be performed effectively and efficiently.

Corporate Recall Planning Procedure—General Guidelines

An important task is for management to provide a Recall Planning Procedure so staff and line personnel have an organizational framework within which to work and a guide describing the responsibility of each person.

Although there may be other objectives set forth in a procedure, the main decisions to delineate are the following:

1. Who is to conduct the recall and who will perform what tasks during the recall?
2. How to protect the customer, user, bystander, and company?
3. How the company will assure compliance with the various government laws and regulations?
4. How to remove, upgrade, or repair unacceptable or questionable products from the marketplace at minimum cost and with minimum inconvenience to the customer?
5. How to protect the assets of the corporation?

The procedure should provide an adequate plan to direct the lead management individual to state the main objectives of the recall. The plan should also direct those individuals involved to

1. Specifically identify the product(s) that are involved.

2. Clearly differentiate between the product that is being recalled and the product that is not subject to recall, repair, or replacement.
3. Identify the location of the product to be recalled.
4. Stop the manufacture of, and further distribution of, defective product.
5. Isolate, correct, or make disposition of nonconforming product(s).
6. Be prepared with a plan of how a specific product would be recalled prior to actually recalling the product.
7. Identify who will do what within an organized response to a recall event.
8. Adjust the product development/manufacturing system to preclude a recurrence of the problem, within acceptable economic and technical limitations.
9. Obtain a clear understanding of the various activities and outline the operation and the duties of the various individuals involved in the activity.
10. Prevent further damage to the organization.

A careful definition of the decisions and who is to perform each of the activities will result in a timely response to the problems, protection of the customers, and a minimization of expenditures. The planning must provide for a system of identifying where various products are shipped and, if possible, their present location. In other words, there is preplanning, which if properly done will minimize the effort of identifying where product subject to recall is actually located. Careful attention to objectives and details can minimize damage to the corporation.

As noted, an important portion of the procedure is the definition of *responsibility*. This may have a double meaning: the responsibility of preparing a system and the responsibility of operating a recall. It should be noted that there are many members of the organization who will have to

1. Participate in conducting a recall
2. Be consulted
3. Be required to review a "Recall Planning Procedure"

The groups typically involved include legal, marketing, finance, manufacturing, quality assurance, distribution, insurance, engineering, service, replacement parts warehouse, and other departments deemed necessary.

The recall plan must define in detail the role that each individual and group in the organization will play. It must also develop the system to fit the particular corporate environment so that action, when needed, can occur promptly, efficiently, and logically.

Alternative Methods

There are several organized alternatives that have been used in a recall environment. Each of these includes the prime objectives of a recall, which are to

1. Protect the customer by timely removal or correction of offending product in the field.
2. Protect the organization by the timely removal or correction of offending material in the field.
3. Provide prompt and economical correction of the problem through a preplanned operation.

Advance Planning

The only way to conduct a recall campaign with minimum cost and damage to the corporation is through careful advance planning.

There are several ways to organize a Recall Activity.

1. *The assignment of total responsibility to a high level executive/manager*. This individual has total responsibility for all product recalls, including planning, training, and organizing for future recalls as they are required. This individual would have the authority to initiate a recall, decide what portion of the product to recall, whether the product is to be reconditioned or destroyed, and whether the product is to be serviced in the field. This system needs adequate communication with other members of the organization and can be more difficult to conduct within a large corporation.

2. *The assignment of a committee to assume the responsibility for all recalls*. The disadvantages of a committee are that it can be difficult to come to a decision. It may make a better decision, however, than an individual. A committee has an advantage in that it spreads the authority through the organization, creating more effective teamwork.

3. *A team approach*. The team resembles a committee except for the fact that the team leader makes the final decisions after consulting with the team members. The advantage of this system, as well as the committee system, is that prior advice to the various departments is provided before the operation begins.

4. In each of the systems, it is necessary that someone in each portion of the operation knows what is required. In some instances a special administrator will need to be appointed.

There are two essential elements of the recall task: a leader with a staff and a procedure.

The leader and staff should be selected from those who have familiarity with the product so they may be able to answer questions about the products and deal effectively with the users, or others who may need to know how to handle the product for return or for service.

The second item needed is a recall procedure. In each system, a method of carrying out the recall operation by each portion of the organization must be set forth so that it can be effectively, expeditiously, and economically performed. The recall procedure should be as specific as can be provided.

One of the best ways to minimize the extent of a recall is to be able to identify the exact groups and products that will have to be recalled. With large items that are identified by serial numbers this will be easier if careful records are kept that identify the components and parts used. In one instance, an automobile manufacturer was able to keep a recall down to a few hundred vehicles with defective bolts due to the adequacy of the records. Other methods include the marking of components with codes or identification marks, preferably automatically, which distinguish the time of manufacture, the lot the pieces came from, the vendor, the material, and other information. In combination with these marking systems are computer records that automatically record the information regarding the characteristics and sources of the items.

Recall Procedure

In developing a recall procedure, each organization must tailor its specific needs in detail to the circumstances that exist at all stages from acquisition of material and parts, through manufacturing, distribution, and use. There may be different procedures for different products that have different fields of operation or widely different characteristics. One company made two types of product, one was sold directly to users, the other sold to other manufacturers. Obviously, the recall procedure and records for these products were grossly different.

Recall procedures should be defined in as specific terms as possible. Assigned authority and the actions of each of the assigned individuals should be listed. Among the step-by-step instructions should be how to decide whether a recall is necessary and to cover each step in the process including the following:

1. Special supplies
2. News releases and notification to owners, users, and distributors
3. Sample form letters
4. Complete check lists of each step to be performed, and who is to do it
5. Government agencies to be notified
6. Correction and repair notices
7. How items in the field are to be identified and located
8. Necessary paperwork, forms, and model notification papers

Chapter II

Finances

Corporate Recall Finances

A second task critical to providing line management with a working framework for a product recall is the providing of "time and expense" rules and their appropriate accounts. A method is needed to capture the costs involved in a recall and allocate the costs to the appropriate operating entities, and placing financial responsibility where it belongs. Since accounting systems in corporations differ and account numbers are different there will be no specific suggestion made here. The item's costs that should be noted, however, include most of those in manufacturing and marketing such as

- Administrative service
- Labor
- Materials

- Facilities
- Insurance
- Legal fees
- Communication
- Goodwill
- Design
- Notification
- Advertising
- Shipment
- Temporary help
- Printing
- Travel

These are not continuing costs, and although unbounded by budgets and estimates, must be carefully scrutinized to avoid inheriting cost center loss leaders.

These costs should be estimated at the planning stage, tracked administratively as the recall proceeds, and revised as needed. Careful and systematic cost accounting is a critical management tool used for planning the recall, administering it in a cost-effective manner, and providing feedback on what has been accomplished and how it relates to projections.

A recall draws on existing in-house facilities: engineering, marketing, and so on. It is important that the recall costs not be assigned to the participating organizations without noting the purpose for which the expenditures were made.

Cost Factors

Administrative costs are "overhead" general costs that continue to be incurred throughout the length of the recall program. They consist of the cost of setting up and administering a record keeping function for the recall, the legal and management costs involved in planning the recall, and dealing with issues that arise as it is implemented.

Tracking and monitoring is a record keeping function. This system can range from a simple pencil-and-paper procedure where a few items are recalled to a complicated computer-based system where the number of items recalled is very large. Costs include any additional personnel assigned to design the system—as well as any additional personnel who maintain and operate the system—those who generate reports and those who keep track of progress.

Legal counsel and management costs cover the in-house attorneys and company management who deal with the contacts with outside counsel, insurance companies, and regulatory agencies. These contacts start with the period the recall is being considered, continue through the period the recall is in progress, and stay in place until there is consensus that the recall is completed or there is agreement that the recall should be shut down.

Management costs must also be built into the financial framework for recall. These costs consist of the time of the recall manager and of those executives at various levels of the firm (for example, marketing, production, quality control, and so on) who have contributed to the design of the recall and are involved as decision makers on issues that arise during the recall.

Labor costs must also be built into the framework of the recall. This is to cover the people who do the actual work on the product and does not include the administrative help who track and monitor the recall.

Material and parts costs must also be maintained. They may include internal and external costs that will be incurred when work is done and the purchase of replacement material in recalled products.

There may also be design costs and service costs. Design costs are incurred when design changes are worked on to prevent the recurrence of defects. Service costs are the costs from outside services for work on repair and reconstituting recalled items. There may also be notification costs: letters, advertising, and news releases.

Sometimes there are insurance costs involved, due to increases the insurer demands, and the service the insurance company may be rendering to expedite the return and recovery of merchandise.

Notification Costs

Notification to product owners is a most important dimension of a recall. A variety of potential tools are available, and they vary considerably in their applicability, cost, and effectiveness depending upon the product situation. Some common principles apply in their choice.

The scope and level of notification efforts should reflect the seriousness of the hazard. Recalled products involving significant hazards will require a more comprehensive media campaign as well as more substantial efforts aimed at *directly* reaching and influencing product owners and end users than will recalls of less hazardous products. For the latter, a limited program of well-designed notices to dealer and distributor networks may be sufficient to discharge reasonable notification efforts.

All notification efforts must be designed to communicate clearly and in a visually acceptable manner that a recall is being conducted, what the hazard is, and what the consumer should do to deal with it (including recommendations as to what constitutes safe product use during the interim pre-remedy period).

All departments in the firm should review draft notification materials for errors and ambiguities before such materials are released.

The mix of recall notification tools chosen by the firm should reflect careful planning, which embodies an assessment of expected cost effectiveness for the tools chosen, and a rationale for rejecting those not chosen. As the recall is in progress, the firm should evaluate which notification tools have been most and least cost effective in reaching and influencing product users. These evaluations should be taken into account in design of subsequent notification efforts.

Chapter III

Traceability

Product Traceability

There are several portions to the traceability function including:

1. *Identifying the component or arrangement that is causing the recall.* This component or arrangement may have been used in the entire production of the product or only a small portion of the product. As an example, a recall may have been caused by a component such as a fastener. In such a situation, it may be wise to examine the sources of fasteners and the lots involved in the products, some of which may be contributors to the problem. If internal records do not indicate which dated products used the troublesome fasteners and there is no way to identify the defective group by sight, the customers would have to return all the items produced and the factory would have to replace all the fasteners, or if the product is to be

repaired in the field, all the fasteners of a specific type and size would need replacement. The size of the recall would then be determined by the total product produced. The difficulty in foreseeing this type of problem and keeping records of where each item went can be substantial. In some instances, records are maintained and the number of units that must be recalled may be small. This may not be a problem in parts that are larger and easily identified.

2. *Identifing where the product that is to be recalled actually was sold and who is the user/owner.* This can also be a formidable problem. With large items such as automobiles, there is a requirement that the manufacturers keep track of where each item was sold and to whom it was sold. On small retail items, this is almost impossible and the only way to identify where items are is through postings, advertisements, radio, news, and similar announcements.

   There is also the difficulty of tracing the combination of what units have the component and where these assemblies are.

   There are those who believe that the manufacturer should have the replacement in the hand of local distributors before a user is notified that there is a problem. This, however, presents some difficulties in that the user/owner may not have been warned promptly.

3. *Identifying where the product is, whether the product is large or few in number, such as a vehicle, and whether there is knowledge as to where the product is.* A vehicle may have to be driven to a recall site. In other cases such as large installed products, the site where the product is must be visited and the correction made. A more common situation is a small retail product with large quantities in the hands of unregistered consumers. In this case, radio, TV, and general publicity through retailers is the only way the product can be located, and recalled. The customer or owner must be given a way to identify the defective items and differenti-

> ate between defective and nondefective items. Repair and/or replacement, although not particularly lengthy for each item, may take considerable time if there are large quantities of items to be serviced. If the repair is to be by the owner, this presents another set of problems. How does one know whether it has been done and whether it has been done correctly?

The problem as to where the product is can be significant. When this occurs, locating the product becomes a major cost item and can delay the completion of the recall for a considerable period of time. This adds to the cost and increases the danger that some untoward incidents may occur.

There is a requirement that action be prompt for many products. This is particularly true if the product is regulated by any of the following:

Food and Drug Administration (FDA)

Consumer Product Safety Commission (CPSC)

Department of Transportation (DOT)

Bureau of Mines (BOM)

Department of Agriculture (USDA)

Federal Aviation Administration (FAA)

National Highway Traffic Safety Administration (NHTSA)

Most of these government agencies require prompt notification in the event that either the risk or severity of injury is great or there has been too many serious occurrences.

The Federal Trade Commission, (FTC), has developed a policy that calls for a joint supervised recall. This may increase the cost and necessity of a recall where the product is truly defective and dangerous.

It may well be that no recall decision is made without the helping hand of a federal agency. Nonetheless, the basic considerations for managements are the same. This again places emphasis on the preplanning and the care necessary to provide a safe product. Since there is always a chance that an agency or management will decide that a recall is appropriate, however, planning and coding product in a way that can identify lots of final product, which contain specific components and parts, may well be economical. This will allow the identification of units containing the part or component that is responsible for a recall. In this manner the recall can be constrained to only a relatively small fraction of the total product made. The saving from a small recall can save more than the entire identification program.

One method of identification is by a code, date, or lot number, which identifies a product lot containing each component or part lot. Identifying lots in this manner may make it possible to recall a minimum number of items when a component is responsible for the recall. Identification may be by code, date, or lot number, and should be marked on the product in some form. Whether this be a serial number or lot number makes little difference if there is a history of what went into each unit. This marking can often be done by a holder, fixture, or jig that holds the item as it is assembled. There is no extra effort required except the historic knowledge of what went into a product by date or code. This type of identification can minimize the number of items that need recall and save large sums of money. Thus the failure of a purchased component, or batch, will result in a far less extensive recall.

Conceivably all of the product believed to contain the purchased "problem component" will need to be recalled. If on the other hand there has been a successful effort to see that each supplier kept records and these were integrated with the companies records of what was assembled into each lot or item, then the recall could be less extensive.

When the risk of recall is significant, specifications for critical components should include a traceability and record retention plan by the assembler of the final product. On the other hand, few if any manufacturers consider the risk likely, and therefore are particularly hard hit by a recall that has no preliminary planning. The record system should be as simple, but as specific and detailed as possible. The retention time of the record system should also be considered. It should not be shorter than the useful life of the product.

When middlemen come between the manufacturer and the ultimate user, the identification of the ultimate buyer is more difficult to determine. This means that lots shipped to distributors should be identified and where possible, the ultimate buyer should be encouraged to return an identification card so that a record can be made of where the product went. Relying on the middleman for record keeping should be encouraged but not completely relied upon. In many industries there is no complete knowledge of where a product went except through the distributor or seller, which complicates the task of a recall.

These factors increase the emphasis that must be placed upon the complete and easy identification of each item in some manner that is permanent and readable by the user after the product is placed in operation. Items such as soft goods, towels, sheets, and similar items are less easily marked, although the use of care and identification tags do provide some areas and methods of traceability. In items where there is a high degree of user risk, this becomes an absolute necessity.

Cost of Traceability

There are several items that contribute to the cost of traceability. These are the in-house costs that a manufacturer incurs by record keeping and identifying its products on an item-by-item or lot-by-lot basis: the cost of finding to whom the defective product was shipped, to whom sold, where the defective product is, and the cost of notifying owners and users of defective product.

In the event that the manufacturer has no internal or product marking, the recall may require the recall of all of the product manufactured and shipped. The cost will include the cost of locating as many items as possible and the cost of recovering, "fixing," and returning each unit.

In the event that the manufacturer has a method of identifying the groups where the defective component was present, it will have a much smaller recall cost.

It is possible that even monthly identification of manufacture will reduce the number of items that must be recalled. The manufacturer must determine the most economical procedure—identification by either lot, date, day, week, year, or individual item (useful on expensive items)—and adopt this procedure. This will minimize record keeping and the traceability of items that cause problems, but even this may not be helpful when a fundamental error causes the entire production run to be offending.

The traceability function that identifies who has the product is a similar story. Different products in a manufacturer's line may need different procedures.

With expensive products, the manufacturer is more apt to know who bought them and where they are. Keeping records and identifying what went into specific units may be easier since products in this category may have name plates and serial numbers, but if no records are kept the problem may be just as bewildering as with less expensive items on which no records and no preplanning had existed.

Factors to Evaluate in a Traceability Program

Each product type is unique, and therefore each traceability program will be unique. The costs may be associated with the product category, product life, and other product characteristics.

Product Category

A manufacturer should identify each product category or type by some unique designation, such as a name or part number. The problems, costs, risks, and criticality of a traceability program are intimately related to the product line and category. Government requirements and reports, historical data, and so on, may be used as guides in the classification system when more than one product is involved.

Variations in a recall could include the following:

- All of a specific product is to be recalled.
- Only certain portions of a product are to be recalled.
- All of a component or all components need to be recalled.
- Only certain lots of components need to be recalled.

If all of the units that have been produced are to be recalled, it should be relatively easy to identify what is to be recalled. If the other situations are the true state, there is need to be sure that good units are not recalled.

The major difficulty is to describe units in such a manner that the owners and users can identify those units that are being recalled and those that are not. The ideal way is to use short blocks of text and illustrations.

Product Life

Product life may be defined as the sum of the shelf life plus the *maximum* useful operating life. This becomes an important consideration in a traceability plan evaluation because of a number of factors.

First, the longer the product life, the greater the number of similar items are likely to be present in the distribution chain and in the hands of users. If the reason for the recall is a nonconformance that existed during a specific manufacturing time period, then many conforming items will be recalled if the nonconforming items cannot be segregated by the users.

Second, over a period of time it may be impossible to determine which design or process revision have been incorporated in specific groups of items.

Third, for longer lived products, the manufacturer and the distributer must retain records for longer periods to provide traceability in the event of a recall. Distributors may have shorter lives than manufacturers. This is a good argument for maintaining the record at the manufacturer's site.

Unit Cost

Unit cost is defined as the sum of the costs of design, manufacture, distribution, and sales. In general, the greater the cost, the greater the incentive to recall or service problem items. This may cause the manufacturer to maintain better traceability.

Recall or Modify in the Field

Some large complex items must be corrected or rectified in the field. This is the most economical way. Sending a technician to a field area or to the actual installation to correct the product and eliminate the problem is one way. Another is the way it is done in the automotive industry. Specific identification and instructions are provided so that a skilled technician can correct the problem at a service center.

Product Complexity

Complex and expensive products such as vehicles, aircraft, and large, not readily moveable products, and products with many components may make it more difficult to maintain a good traceability program. The problem is the complexity and the large number of components, not their location. In some instances there are few manufactured, which makes for greater ease in traceability. In some items it is necessary to keep accurate track of even the smallest components.

Level of Downstream Traceability (Reason for Recall)

When there has been difficulty with the product or there is risk that the user may have difficulty, a search of the various operating and misuse modes should be explored. Is the reason for having a recall due to a manufacturing process, a design problem, or due to installation or associated with the product operation? A list of potential failure modes may be determined from a Safety Hazard, Failure Mode or Fault Tree Analysis.

If it is a manufacturing process problem, the manufacturing codes and dates, lot identification, and facilities are necessary factors to be analyzed. If there is a design safety defect that is responsible for the recall, the model, revision, and serial numbers may be the important factors to identify.

For each failure mode, a determination will have to be made concerning the level of traceability required to segregate the defective items to be recalled from the total population.

Documents Providing Traceability

Based on the analysis of failure modes the necessary type of information that will identify and differentiate between the items to be recalled and the rest of the population should be determined. Among the useful information may be test records, a list of critical processes and their performance records, date of manufacture, supplier lot records, lot numbers, shipping records, and so on.

Type of Identification

Each item of product should be identified at the lowest level economically possible in order to maintain identification of like items. Serial numbers may be costly if not automatically retrieved by computer. These however are usually assigned to high-priced items. Dating is one method. Bar code type markers—which can be applied automatically and list each item, part, or component as it is released and identify the item into which it was assembled—are a newer method of identification. In some aircraft assembly each item is recorded as it is assembled into the aircraft and records and names of all replacements are kept by manufacturer, type, and serial number. As in drawing systems, product identification may include model number, serial number, date code, and change letter.

Coded Versus Uncoded Identification

The decision to code or identify dates by code is a problem that may affect sales. If it is apt to have an adverse effect, then code the date. Uncoded information is easier for the customer to understand and helps identify the proper item to return or get serviced in the event of a recall. Government regulation will sometimes demand that uncoded information be furnished. This means that one should always be aware of the regulations that apply to the product or if there are any.

Method of Identification

Many physical products can be identified by a name, tag, name-plate with a ink stamp, or indent. Many soft goods items are more difficult to identify. Some can be identified by labels as in shirts, coats, and dresses. It is important that permanency be considered.

Chapter IV

Records

The Record Required for Traceability

Locating the Product

When the need for a product recall becomes a reality, an effort must be made to locate the product that must be recalled.

- The product may be in inventory.
- The product may have been sold to distributors, jobbers, or other marketing "arms" of the producer.
- The product may be in the hand of the final user.

The generation of records to identify the location of these portions of the product must be planned prior to the time shipments begin if this is to be easily performed.

The operational challenge called recall, forces the manufacturer of a product—particularly one that may come under the aegis of a governmental agency—to be sure that the product, procedures, and record keeping will give him the maximum assistance in conducting a recall of product if the necessity arises.

The following are some of the records to be considered to aid in the identification and traceability of the product under recall

- Manufacturing data and records
- Test records
- Shipping and distribution records

Manufacturing Data and Records

Parts lists, engineering changes, product configuration control information, raw material supplier lists, and component and subassembly supplier lists are examples of manufacturing data. The effect of different suppliers of materials and each supplier of parts may identify a small quantity of product rather than the product made over a long period of time and thereby reduce the quantity of product that must be recalled.

When engineering changes or other changes are incorporated into a product, it is worthwhile to retain the time and identity of product that has been manufactured or shipped before and after the change. This may help identify the product that must be recalled. Obviously, there should be some method of identifying product before and after the changes.

The engineering, manufacturing, and other divisions of the organization should cooperate to maximize the efficiency of the record keeping, the marking, and identification of changes in the product. There are often areas in the manufacturing system where material is held. A method of marking can be inserted at this point. This has minor effect on the increase in cost. In other instances—aircraft and motor vehicles—the records are more exact and can be associated with model and serial number, or marked by automatic equipment.

Test Records

Records of tests performed on product at various periods and before and after design changes should be formalized, signed, and dated. This type of information can be effective in determining differences in product that can be reviewed at the time a recall is projected and help identify groups that should and need not be recalled. This information should be included with the shipping date, code, serial number, or lot identification so that groups of product can be recalled rather than the entire manufacturing group.

Shipping and Distribution Records

When a recall is required, the knowledge of where specific lots of product were shipped and through what system they were distributed may be of great importance. Specific groups may have gone to certain areas or may have been distributed by specific distributors. This may simplify a recall. In initiating a recall important information concerning

1. Which product, by type, design, manufacturing period, or with specific parts, components or sources must be recalled.

2. How much of the product is in users' possession and is to be recalled. An estimate can be obtained from the following formula:

   $U = T - (I\&S + D + J + R)$

   U = Product in users' possession

T = Total suspected product produced

I&S = Suspected product in Inventory and Shipping

D = Suspected product at Distributors

J = Suspected product at Jobbers

R = Suspected product at Retailers

Determining the quantity of suspect product and obtaining its return from distributors, jobbers, and retailers may be much easier than determining the location and obtaining the recall of suspected product from the end user or customer.

The records maintained on items shipped to distributors, jobbers, and retailers, and the use of warranty cards or other forms of user information may greatly simplify the recall requirement.

There may be a number of conditions to be considered:

Is the product to be recalled and reworked?

Is the product to be recalled and replaced?

Is the recalled product going to be destroyed?

Is the product going to be reworked or reconditioned in the field at specific sites, as at dealers sites, or on site if it can not be moved?

There may be a difference between defective product and suspected product that must be examined during a recall. Is this product in inventory, in the hand of distributors and jobbers, the hand of retailers, or in the hand of the final owner and user? If the product is consumable, how will we identify that which has been used and that which is still in service or in stock at various locations.

This latter is one of the perplexing problems that face a recall.

Before a recall action is implemented the company must consider how it wants to dispose of the recalled items, or whether disposal is the proper procedure. A change in labeling accompanying the product is technically a recall, yet it does not result in a returned product or a disposal. The traditional ways of disposing of the recalled product is either through destruction or rework and reincorporating the product into product for sale or return to the customer. The activity may take place at the customer's, distributor's, or manufacturer's site, as well as that of a third party.

When the recall is planned, the planner must consider the nature of the defect (for example, does the company have enough expertise and manpower to correct the problem itself, contract it out, salvage some or all material, or rework it into other or subsequent product?). Some products pose a problem if disposed of improperly (for example, batteries and refrigerators). What are the legal requirements concerning the disposition of such products (FD&C Act, environmental regulations)? The economic factors associated with the disposition of the recalled product should also be taken into account. There is also the problem of whether the material can be recycled (or must be recycled). This has become a major problem with plastic, metal, and paper products.

Chapter V

Review and Retention of Records

Review of Records

An organization marketing a product should periodically review its own records to determine if the desired results are obtained by the record system. Some of the questions that should be asked if a recall must be initiated are the following:

1. Is there a way to identify the product by date or short period of manufacture, by a given configuration, by a block of serial numbers, or by some other means of identification on the product?

2. If the product was produced at more than one location, can the suspect location be identified by the user/owner?

3. If the product problem is associated with one of a multiple molding or casting procedures, is there some identification on the moldings or castings that the user can identify?

4. Is there an identification procedure that will make it possible for the owner/user to identify the product with some test procedure report, period of manufacture, design change, or date code?

5. Is a first-in first-out (FIFO) plan used in shipping product from inventory and storage?

6. Are accurate records maintained as to shipment made to distributors, jobbers, and retailers?

7. Are returned warranty cards or records kept and properly stored? Can these records be associated with the date of manufacture, the site of manufacture, the product manufacturing method and components, and/or serial number?

Computer Records

It should be noted that the use of computer records makes it possible to obtain mailing labels and locations of users with much greater ease than by copying from returned cards on a manual basis. The storing of records should be based on the possible need to rapidly gain access to records of user addresses, even though these may have changed over time.

The services of an expert on information retrieval may be necessary and may be a worthwhile investment before the material is preserved.

Retention of Records

The length of time records should be retained is recommended by the following liability calendar:

Liability Calendar

Design—development phase, estimated average	3 years
Manufacturing period—estimated maximum	10 years
Distribution delays—estimated maximum	5 years
Period of use before incident—estimated average	6 years
Loss, Injury, Damage, or other incident	
Statute of limitations—nominal	3 years
Trial delays—nominal	3 years
Appeal delays—if applicable	2 years
Retrial delays—if directed	2 years
Total data requirements	34 years

If a minor is involved in an incident, the time until he reaches maturity may become part of this holding time and further extend the time. A difficulty becomes apparent. There may no longer be personnel in the organization who have any familiarity with the product. Therefore the material preserved must be self explanatory and more than just notes.

In addition the data must be preserved in a manner that is recoverable. Some forms of data, even those on computers, are extremely difficult to interpret or read after the period of time listed.

Many have noted that in the event one or more incidents have occurred, the legal staffs of both organizations involved in the litigation will have filed suit and this will provide an opportunity for the identification of documents that must be preserved. In other situations, most organizations have procedures where material is discarded after a more reasonable period of time.

Chapter VI

Keeping Up with the Law: New Requirements, New Procedures

Once a system has been established to formulate, obtain, and retain the records necessary to plan and implement a recall, an organization should not sit back and assume that the system established is complete and operable for all time. With changes in products, production methods, marketing, distribution, the law, and the techniques for retention and retrieval of records, the systems should be periodically reviewed. Frequency of the reviews will depend upon changes in products, management and system methods, as well as changes in the law. These reviews may determine errors and misinterpretations, as well as better and more efficient methods of maintaining the data.

The foregoing discussion on records required and records retention is based on a product that flows through the entire system; that is from manufacturer through distributors, jobbers, and retailers, to the final user. The need to maintain identification, traceability, and records is just as important for suppliers of raw material, component parts, subassemblies, as to producers that sell to original equipment manufacturer's (OEM) markets.

In case of recall or in product liability suits, the suppliers may be brought into the procedure as being partially or wholly liable.

Joseph Hutchinson, in *Product Recall: Procedural Considerations*, said, "If you do have a recall and you do handle it efficiently, it is just possible that you may be able to prove to your customer that, even though a mistake has been made, you aim to correct it and your company stands behind the product."

Part II

The Recall Process

Chapter VII

Decision to Recall—Performing a Recall Operation

Is a Recall Justified?

Recalls are initiated to reduce the frequency of accidents and injuries, to reduce exposure to liability, to improve performance, and to conform products to specific laws. Many products come under specific laws. Among these laws are the Flammable Fabrics Act, the Consumer Products Safety Act, the Federal Hazardous Substance Act, the Poison Prevention Packaging Act, and the National Transportation Safety Board's regulations of aircraft, automobiles, trucks, and other transportation products.

Could your product require a recall? The list of product recalls is endless. It includes such things as food products, drugs, heart pacemakers, other medical products, toys, electrical appliances, automobiles, automotive products, power tools, smoke detectors, other safety devices, watercraft, housewares, kitchen appliances, exercisers, nursery furniture, garden equipment, textiles, clothing, communication devices, liquids, gases, manufacturing tools and lines, prostheses, building components, plumbing products, chemicals such as pesticides, construction equipment, construction materials, paints, and finishes.

Any time after a product has left the manufacturing site and it is determined to present a mechanical, electrical, chemical, thermal, instructional, ingestion, or other hazard, it may be subject to a recall. A recall may be for replacement, repair or correction, or retirement of the product. The judgement of the seriousness and/or the consequences of the hazard may be made by the manufacturer, the media, or by a governmental agency. The determination that a hazard exists may be in dispute. The manufacturer may maintain that no hazard exists, and a governmental agency may insist that there is a hazard. Situations such as this must be resolved to the best interest of the user.

It is always prudent to remember that the best interests of any business are inexorably tied to the best interests of its customers.

Consideration

An organization, in many instances, must make a decision that is in *its* best interest—to make a recall—and from there on make every effort to be thorough and protective of its customers and itself.

Decision Basis

An important part of the decision process is the determination of who is exposed to a hazard presented by the product. These may include

- Those who produce, test, or sell the product
- Those who handle and transport the product

- Those who use or assemble the product
- Those who maintain, service, repair, or dispose of the product
- Third party bystanders
- Those exposed to harmful waste products
- Those exposed to vapors or other outputs of the product

The people exposed to harmful effects will have a bearing on whether a recall is initiated. If, for example, the group exposed consists of those who produce, test, or sell the product, it is unlikely that a recall would be required. Corrective action would be required to protect the members of the group affected. What that procedure might be would have to be determined by the producer and seller.

The nature of the possible injury and the number of people who might be exposed to that injury are two important considerations that determine whether there should be a recall. Another factor is the possibility of damage to equipment or property.

Chapter VIII

Implementing and Evaluating a Recall

Initial Activities to Implement a Recall

A Management Control Center should be established prior to a recall decision. A management product safety committee should be set up under the overall recall plan in order to operate and monitor the total program.

This group should be representative of the internal departments directly concerned, such as Engineering, Quality Control, Manufacturing, Legal Service, Insurance, Consumer Affairs, Marketing, and so on, for the particular service in question. This managing group can either be the same as or a delegated activity for the recall manager or committee.

The managing group of the committee should prepare a notification containing information about the product, its features, and the reason for the recall. The notification must be in compliance with any government rules and regulations that apply.

This management control group should establish standardized reporting forms and procedures for every aspect of the recall. This standardization should include means by which summary information can easily be presented to management from time-to-time for their evaluation of the completeness of the recall. The submission of such information to an interested federal agency would be the responsibility of this group.

Periodic reports and reviews should be made by the financial people so that costs can be properly identified and controlled. Reporting and analysis documents should be the same (or similar) to records regularly used within the organization.

Description of the Product

The company involved in a product recall should clearly and specifically describe the product, the problem, and nature of the hazards involved. In addition, the potential risk involved—including all possible effects on the consumer—should be included. If required, proper notification could be given to any government agency, actual consumers or users, and others who might require a clear understanding and information to properly react to the product and the recall.

Identification of Products

Clear and unique product model identification should be established for all products, and used in the effort to obtain the return of the defective product that is in need of action.

Due to the many variations of product design and classification, each particular product should be clearly described including generic name, manufacturer, date manufactured, type, company brand name, model numbers, serial numbers, and all other identifications on the product. In the event of a recall, the product must be clearly and specifically identifiable by the consumer and field service personnel.

Also any additional physical information associated with the product in question should be included, such as color, size, material, overall physical structure, and so on. This product information should be standardized, available at any time, and permanently retained in at least engineering, marketing, or manufacturing files.

Definition of Product Groups to Be Recalled

The entire identifying data such as model number, serial number, manufacturing dates, and so on, should be easily identifiable and clearly noted in all information released, and the range of such numerical sequences should be specifically outlined. Any prefixes or suffixes should be clearly noted to define the particularly affected product in the field. The product labeling design should clearly include such information on an easily accessible marking format. In this way all information being released to the consumer and any other public relations media is made known to all persons concerned.

Data on Population of Products

The overall product population involved in a recall should be easy to trace and identify for each variation of product model or design, as well as the total number of product units actually existing in the field at the time of the recall. Key marketing, production, and distribution personnel and their primary subordinates should maintain product data accurately and in an easily accessible form.

Such data are necessary to identify the variations in the population that may exist for the product, such as units within the manufacturer's own inventory, at the distribution level, at the retail level, and the number of product units in consumers' hands.

Accurate records of this population will provide a measure of the success of the recall.

There are variations in recalls. These include

- When all of a specific product is to be recalled
- When only certain portions of a product are to be recalled
- When only certain lots of components require recall

If all of the units that have been produced are to be recalled, it should be relatively easy to identify what is to be recalled.

When a portion of the product should be recalled, the questions that arise are "How can the manufacturer, distributor, or customer identify the units that are to be recalled and differentiate them from those that are not to be recalled?" and "How can a small portion of the product that must be recalled due to the inclusion of a problematic procedure or component be identified?"

The salient points to provide the access to these questions are

1. An accurate record of the end product in which each component lot was used
2. A record of the end items

It is therefore desirable that the product have an identifier that changes with each change of the component, part, production run, and any other change. These records add costs but the savings at the time of recall are more than sufficient to make up the difference.

Proper identification throughout the process of manufacturing, identifying the components contained, and their sources as they are manufactured can assist in reducing the quantity of product that must be included in the recall. This care in processing can pay off multifold by reducing the cost in the event of a recall.

Internal Communication

In a large corporation it may be difficult and time consuming to get information to the proper management personnel. In a potential recall situation, time is of the essence. To ensure prompt communications, field personnel (especially sales and service representatives) must be trained, and notification procedures developed to alert all appropriate personnel to a possible recall.

Chapter IX

Cost Considerations

Recall Costing

A recall may be expensive, but it may be a lot less expensive than the cost due to injuries and losses. Nonetheless, a critical aspect of a recall is the accounting of its costs. This involves both estimating these costs at the planning stage of the recall, tracking them administratively as the recall proceeds, and revising them as needed. Careful and systematic cost accounting is a critical management tool for planning the recall, administering it in a cost-effective manner, and providing feedback on what has been accomplished and how it relates to projections. Such cost accounting begins with developing a recall planning budget and budget categories, which then constitute the framework for cost-effective managing, monitoring, and tracking the progress of the recall over time.

The difficulty with recall costing is that in most firms a recall is special, rather than a regular recurring organizational event. Thus many organizations have never used accounting protocols, which formally provide for recall costing. Furthermore, since resources for a recall are frequently drawn from numerous and diverse cost centers within the firms, significant cost allocation problems exist that must be surmounted. Illustratively, a recall draws on existing in-house engineering, marketing, distribution, sales, and legal resources of a firm as well as physical plant, overhead, and general/administrative resources. These resources are reflected financially within well-defined cost categories corresponding to existing organizational structures financially. When a recall mobilizes these existing resources into a new activity for accounting purposes, problems arise in allocating the financial costs of these resources between the recall and the regular activities in the firm. The firm's accounting and financial management staff is usually in the best position to resolve the difficulties involved in recall cost accounting.

There are numerous potential cost categories that may be important in a recall.

The following represent some of the most salient cost considerations. They include administrative costs, tracking and monitoring costs, legal costs, management costs, labor costs, material costs, recovery costs, and reward/bounty/incentive costs.

Cost Factors

"Overhead" General Costs

Overhead general costs are costs that continue to be incurred throughout the length of the recall program. They consist of the cost of setting up and administering, monitoring, tracking and record keeping function for the recall, and the legal and management costs involved in planning the recall, and dealing with issues arising as it is implemented.

Tracking and Monitoring

A tracking and monitoring system to "keep score" of the recall as it progresses is essential. The system tracks the cost of setting up and maintaining records for monitoring the progress of the recall, and tracks items from the stage of attempted identification of owners or sellers through the stage of getting them remedied under the recall. The system can range from a simple pencil-and-paper tracking procedure (for example, for a recall involving a few products which are in identifiable owner's hands) to a complex, highly computerized system that records plans to reach large national dealership networks and millions of end users. The system must be able to generate reports that measure effectiveness of management's use and provide feedback as to potential elements of the recall that may need revision. Costs include personnel who design the system, as well as those who maintain and administer it through functions such as data entry and report generation.

Legal Counsel Costs

"Product liability counsel" must be employed in the design stage of the recall as well as its implementation phase as legal issues arise involving the interested parties, such as insurance companies, regulatory agencies, component manufacturers or distributors. The attorney may also provide advice at the outset on the need for a recall, and at the conclusion on issues such as appropriateness of closing it out or scaling it down.

Management Costs

Management costs must also be built into the financial framework for the recall. The cost of the time of the recall manager, and of those corporate executives at various levels of the firm (for example, marketing, production, quality control) who have contributed to the design of the recall and are involved as decision makers on issues that arise during conduct of the recall.

Labor Costs

Labor costs include cost for workers who produce the necessary component parts for the recall, perform assembly and parts replacement, answer telephones, deal with mail, assemble and package materials, and so on. It does not include the administrative personnel who track and monitor the progress of the recall.

Material or Parts Costs

Material or parts required to implement a recall, whether manufactured internally or purchased by the firm, often carry with them costs problems. The material or parts may come only at premium prices because of availability and/or schedule problems in the face of timely need. Suppliers or contractors may have to work overtime—adjusting schedules to provide parts or materials on very short notice. These extra efforts have the potential of raising a firm's recall costs significantly.

Reward/Bounty/Incentive Costs

A retrieval premium is infrequently employed in product recalls, but where used may be very effective. It is most common for products involving significant hazard potential where owners may be hard to locate or persuade to cooperate without extra compensation.

Miscellaneous Costs

There are many items that should be included in the total costing scheme. Some of the more salient are facilities costs, design costs, service costs, notification costs, media costs, point of sale notices, direct notices, repair shops, distributors, Internet, consumer clubs, trade material (journals), shipping, transportation and other distribution costs, travel costs for notifiers, repair personnel and customers, cost of getting the item to and from the consumer, and testing and product depreciation costs.

All items that can be substantial in a recall should be separately costed.

Chapter X

Document Retention

All documents associated with a recall should be retained. The tendency in a recall is to assume that once a recall is complete, there is no further need for the documents involved in the recall. This is not so. Recalls may be the subject of inquiries by government agencies or private lawsuits. If a question is raised about the adequacy of the recall, it is usually dependant on documents for an answer.

As discussed throughout this guide, the management of a recall is, in most cases, an *ad hoc* activity. This means that the task will usually be handled by people who are not guided by established procedures. A recall manager should ensure that descriptive and procedural documents involved in a recall are retained for future use. Without the documents, future inquiries will have to depend solely on the memory of people who may have left the company or are difficult to locate. In other words, if something was not written or the writing is not preserved, the event may just as well never have occurred. Thus document retention is an important element of a recall. Some specific documents are discussed in the following sections.

Length of Time

The length of time for retaining key materials should be set after consultation with the legal and product specialist of the company. General guidelines are usually the length of life of the affected product, relevant statutes of limitation on legal actions, insurance issues, and other concerns.

Advertising Retention

If the recall campaign involves the placement of advertising, paid or otherwise, copies of these advertisements should be retained. Also a list of the publications or places in which the advertisements appeared should be preserved. If a clipping service is used to cover the recall, the results of the clipping service's work should be retained to demonstrate the breadth of circulation of the advertisements.

Mailing List Retention

If mailing lists, either the company's or purchased from outside sources, are used in the recall, these lists should be preserved. Further, any record of returned mail, completed mail, or other contacts involving the mailing should be preserved.

Drawings, Retention

If drawings, photographs, or pictorial representations are used in notices, posters, or advertisements, or in the production of a new product used as replacement, these should be retained. If a number of drafts were used to reach the final drawing, at least the final drawing should be preserved.

Manuals, Retention

Manuals distributed either with the replacement product or as some other element of the recall should be retained. If possible, the changes introduced in the new manual should be indicated in the retained records.

Labels, Retention

Labels are a particular point of inquiry in modern product liability theory. In some recalls, the point of recall will be the provision of a new label. If this is the case, the new label, in the form in which it was provided to the user, should be retained. If any studies were done, the drafting of the label, testing of it, or other evaluations should be preserved.

If the object of the recall is the provision of a new product and that product has a label, the label should be retained, both on the product and in the form in which it was applied to the product.

Supplier's Documents and Transitional Support Documents, Retention

Purchase orders, invoices, and other documents used to produce new parts, labels, manuals, or other components of the recall, should be retained. They are evidence of the amounts of goods procured, cost of these goods, and the timeliness of deliveries.

All of these elements are subject to examination in the event of government inquiry or private lawsuits delving into the adequacy of a recall.

Shipping Documents, Retention

In many instances, a recall will call for replacement products or parts to be sent to owners, users, distributors, or retailers. The shipping documents are necessary links to prove that delivery of the replacement product or part actually occurred. In some lawsuits people have claimed that they were not informed of the recall or did not receive the necessary replacement product or parts. The shipping document is an important way to establish delivery.

Press Releases, Retention

If a press release is used to announce a recall, this document should be preserved. Again as with advertising, if any clipping service or other means is used to record or measure exposures for the press releases, these should also be preserved.

Correspondence, Retention

Correspondence, particularly that with a relevant government agency, should be retained. If a government agency deals with the company on this recall, this correspondence can establish the acceptability of the recall. This can be particularly helpful in the case of an agency respected as being an expert in the field.

Chapter XI

Reporting Forms

Planning and Designing Reporting Forms

The Management Product Safety Committee must decide what information is appropriate and necessary to be recorded for optimum feedback from a recall. This information may vary from product type, plant, location, and so on, and should be thoroughly researched in the planning stage. After a recall is in progress, it is *too late* to decide that certain information would be helpful or necessary.

It may be appropriate to provide easily reported procedural information to field management through the use of summary diagrams showing the flow of recall information and retention of these reports in the field. This will also serve for easy reference to lower field management levels, including those who will handle reports and deal with customers.

Reporting forms should be designed to obtain the maximum usable information with the least effort on the part of the consumer and the service department.

Reporting forms, which have been designed for the recall, should be supplied in necessary and sufficient quantities to assure that forms are available at service locations and for consumers. All paperwork and forms should be provided together, not separately, with direct procedures and instruction on how they are to be filled out and whether there may be modifications or replacement reports.

Sending out the reporting paperwork should be planned for as early as possible during the recall program so that the information will reach the field service level at about the same time as any overall modifications instructions, kits, and recall bulletins might be distributed.

A firm initiating a return and refund program should plan to retrieve the appropriate consumer information, specific returned product identification data (for example, Model No., Serial No., version, color, and so on). This will help lessen the exchange and handling of incorrectly returned products of similar types and models. Advice about the consequences of incorrect merchandise should also be included to inform the consumer early and directly.

A brief and easy format should be used in developing the paperwork for service or replacement tickets which must be filled out in the field. These forms should include returned product information.

The raw data should be sent from the field to one person designated by the Management Product Safety Committee who will accumulate these data and submit timely reports.

Release of Recall and Notification Information

A notification must be prepared containing information about the product, its description, its features, and the reason for the recall. The notification must be in compliance with any government rules and regulations which apply.

The notification should also describe the actions to be taken by all people in the distribution cycle (distributors, customers, and users). These actions may include

1. Supplying a toll-free number to be called for further information.
2. Customer or user contact with the distribution center, if appropriate.
3. Instructions as to where the product is to be returned or where the manufacturer or service center is located.

Special News Releases

The information regarding the product and its hazards should be clearly outlined in *all* outgoing releases. In addition, prior planning for any questions regarding the product's problems should be done in order to develop and provide the basis for proper, accurate, and consistent responses from the company.

Communications of product information to public media must be included in the company's plan and the timing (and the reasons for the timing) of the recall. Releases may be sent out by any of the following:

- Special news releases (radio, TV, newspapers)
- Advertisements
- Direct contact
- Internet
- E-mail

The choice may vary depending upon the product, the urgency, and the geographical distribution of the product and users.

Internal Employee Information

Management should quickly and consistently distribute recall information internally so that employees are fully and properly aware of the situation and the product involved. Special bulletins, company meetings, and even a special office for information should be considered, in order that employees can obtain information about the product and its potential hazards as well as the return of units which they themselves may have.

Timing

In attempting to achieve an optimal saturation of its product recall notices, a company should plan the timing of releases in the various media.

Although generally timing is of the essence in implementing a recall, a smooth and efficient process must not be sacrificed. A balance between the level of hazard presented by the suspect products, and the availability of replacement products or parts of the replacement products must be evaluated and effectively determined and then recall implementation must be coordinated with the most effective means of delivering notice of the recall. Some products will be most effectively recalled by notice in popular or trade publications. This requires that the publishing schedules of the appropriate publications be determined. Another example of setting dates has to do with seasonal products. For example Christmas lights, ornaments, and decorations are the focus of retailers and consumers for a specific period of time. Any attempt to recall Christmas tree lights in July would be less successful than a similar attempt in December. Similar timing applies to other products.

A useful means of establishing calendar dates is to determine the amount of time necessary to obtain replacement products or parts and the lead time necessary to get notices to owners or users and then plan backwards from those dates. If a regulatory agency is involved, that agency should be informed of the company's plan and the timing (and the reasons for the timing) of the recall.

Chapter XII

Field Response System

Objective

The primary objective of a recall is to retrieve the product in question. The response procedures to be followed by the consumer or any service group should be clearly spelled out.

Once the specific plan has been decided by the Management Product Safety Committee, and the clearer and more understandable the communication is, the more likely the message will be understood and proper action taken by the consumer or the service group.

A recall is an ongoing activity; as such, the recall manager must monitor what is happening in the field. This includes the response from the company's representatives in the field (for example, sales people, service personnel), other people engaged in the distribution of the product, and the user. Also, in some situations, regulatory agencies may be involved in monitoring the recall.

Response from the Field

To accomplish this monitoring activity, the product's sales, service, and use patterns should be understood. With this knowledge, a useful monitoring system can be established. By monitoring information on the recall, the recall manager can determine if the recall message, delivery system, or any other step in the recall process needs adjustment. Whatever system is developed to monitor field response, it should be capable of memorializing the response and identifying areas for further investigation.

The use of a toll-free number or other means of direct access allows knowledgeable personnel to be available to answer questions that are not anticipated. Sometimes users will have little or no understanding of the recall, but have the telephone number of the company. These contacts must be handled well or they will be lost to the recall effort.

Feedback for Analysis

A vital part of the recall monitoring system is the provision for feedback from the field. Capturing this feedback enables the recall manager to determine if the recall message is perceived accurately by the target audience, if targeted users are responding and if the delivery system is functioning properly. In some larger recalls, the company may want to conduct an *effectiveness study* of its recall. This type of study is useful in situations where questions may be raised about the recall's rate of return. Such studies can highlight further areas for effort or demonstrate that further efforts would not be cost effective.

The level of detail monitoring depends, again, on the type of product, number of units in use, and similar considerations. This information can be very helpful in refining the ongoing recall effort and, if necessary, in planning better recall efforts in the future.

Collection of Internal and Field Cost Information

Every recall has a cost, whether it is merely for field repair, providing parts to the consumer, or a major correction or replacement of heavy equipment type products. To minimize these costs, it is necessary for the groups involved to plan a review of the entire recall, and to provide replacement or correction procedures.

Practice sessions should be held not only for a review of the work time for any modification but also to familiarize instructors who will be advising field personnel how to correct the problem. During these practice sessions, times should be estimated to establish a base cost for modifications including factors for fatigue, learning curve efficiency, and so on.

A standardized cost for a field modification should therefore be planned for and include a worst case situation so that a unit cost can be associated with all field units.

Depending on the type of product recall or modification, costs are incurred for on-site modification, labor, travel, parts shipments, special delivery, and so on. It is important that all such costs be considered in making the total cost estimate since many may be significant.

Consideration should be given to the effect of the recall on fixed expense items. In addition, the total costs should include items such as media fees, expense for notification, travel, communication, special shipping expenses, special legal fees, technical and nontechnical consultant fees, purchase of advertising space, direct mailing costs, and other related costs.

A system of accounting to accumulate each of the costs should be part of the overall plan.

Parts Use Traceability

A recall always has time constraints. In some situations the "fix" is later discovered to have problems of its own. Thus, as products and parts are replaced in the field, adequate records should be maintained in order to trace these products or parts so that subsequent work or modification can, if necessary, be performed.

Field Data Reporting

The information, positive and negative, received from the field should be sent to a single person or group which has the duty to accumulate these data. That person or group can then analyze the data, provide the data to others for analysis, publish reports, and/or preserve the records of the recall's field experience.

The information gathering system for the recall should generate reports periodically to inform the recall manager and company management of the status of the recall. The recall's progress may be measured against time, number of products recalled, geographic penetration, contacts with owners/users, and/or other criteria. Depending on the size and nature of the recall, some or all of these data will be required on a periodic basis to monitor the recall.

Company management may be content to review progress in terms of numbers of products or units recalled or serviced, whereas the recall manager may require more detailed information. Unusual activity or problems should merit special reports, highlighting any aberrational activity. Distribution of these reports should be to members of the Management Product Safety Committee and the firm's executive personnel. In some organizations the distribution should be limited to key personnel in order to avoid possible misuse and misinterpretation of the total program and its achievements.

Monitoring Activity

Monitoring activity of a recall will usually be conducted by the quality assurance function of the manufacturer of the recalled product, a third party certification organization, and/or a government regulatory organization.

Auditing of Field Corrective Action

The monitoring organization should perform an audit of the actual retrofit being performed in the field. This should also be done for material corrected or produced in a factory to assure that the intended correction of the problem has been achieved. In the case of a large population of products requiring retrofit, a representative sample of field corrective actions should be audited throughout the retrofit program. The audit should include verification that

1. The correct replacement parts were used.
2. The proper documentation/specifications were available.
3. The components were installed correctly.
4. The replaced discrepant component is isolated, properly disposed, and/or returned to the manufacturer according to the directions of the retrofit/recall program.
5. The corrected product functions properly after the retrofit.
6. The correct number of completed units and other pertinent data is recorded on the field data reporting form and of the ease of repair.

The manufacturer must be notified immediately of any discrepancies/problems found during the audit.

Evaluation of the Ease of Repair

Consideration should be given to the ease of repair of a product involved in a field retrofit program. The audits of the field corrective action will reveal any problems associated with accessibility, environmental factors, parts availability, special tooling, adequacy, and clarity of technical specifications, training issues and documentation. The manufacturer should be immediately notified of any problems found regarding the ease of repairs.

Data Verification and Correlation

The monitoring organization should verify that the correct number of completed units and the pertinent data are properly recorded on the field data reporting form during the audit of the field corrective action.

Retrofit completion data collected during the audits of the field corrective action should correlate with the manufacturer's summary reports of the recall/retrofit progress. The manufacturer should investigate any discrepancies between the manufacturer's recall summary reports and completion data collected from the monitoring organization's audits.

Chapter XIII

Management Evaluation/Program Completion

Periodic Program Review

As with any other program, a recall program requires frequent and periodic review. During the initial stages of the recall the Management Product Safety Committee should hold frequent meetings and support the operation by reviewing the planning and implementation of the program.

Thereafter, frequent but regularly scheduled meetings should be held to review progress. If the recalled product is a low-volume, specialized item there will have to be a review of the success in contacting the customers and doing the field conversions. For a high-volume, widely used item the emphasis will have to be on the availability of replacement items and the success in identifying the location of product in the field and in the success of the retrieval of the items.

A major interest of the management review is the relative success of the program including

1. Internal effectiveness of the recall procedures and providing company information
2. Evaluation of field effectiveness for return or product modification

Evaluation of the internal effects should include both the hardware aspect of a recall program as well as any side effects throughout the firm's activities.

A product recall program may have various effects on a company. It will help bring to light the overall efficiency of the existing system, the capability of the product traceability methods, correct record keeping and accounting procedures, and the methods of disseminating information throughout the organization.

A recall need not tarnish the reputation of the firm. It may be received by commercial and consumer groups without opposition. This viewpoint will help to achieve a higher level of success and decrease the long-term negative effect of a recall.

Success Level Evaluation

The originally defined product population should be examined in a realistic way to estimate the level of success over both short and extended periods of time.

Factors that affect the ability to locate one's product in the field include the price, life expectancy, regular or seasonal usage, secondary users, and physical environment for use and storage. These items as well as repair and service experience will affect the percentage of product that will ultimately be located and recalled.

Each individual firm and its management must evaluate its own situation to review the effect of its initial efforts and the possible need for improvement to retrieve more product. This is true not only for instances where there is a regulatory agency involved but also from the firm's interest to protect the public from injury or loss and for its own protection to maintain freedom from litigation.

Chapter XIV

Legal Implications of Product Recall

In today's society the public expects corporations to fulfill their responsibilities beyond the financial desires of the stockholder and stakeholder of the organization. Public opinion, coupled with a committed desire by most corporate leaders for a safe and reliable product has set the guidelines for today's corporate/community relationship. The relationship should be sensitive to the needs and desires of the general public and to the raising of the quality level of life by designing and manufacturing products developed to protect human resources. The consequences of ignoring these moral and ethical issues are manifold. Adverse publicity, product litigation, loss of reputation and goodwill, potential risk of legislation, increased governmental regulation, and judicial action are all real possibilities.

There are various judicial actions that are possible against an organization and its management members when defective product causes death, injury, property damage, or economic loss.

From the company's viewpoint, the exposure is to individual or class-action suits. This exposure to seek recompense due to damage is based on the theories of negligent conduct, fraud, misrepresentation, breach of express or implied contract warranties, and strict liability now adopted by almost all jurisdictions pursuant to Section 402A of the Restatement of Torts. Even a civil complaint may include a claim for the award of punitive damages. The liability implications are severe. Nevertheless recent actions have illustrated such exposure appears to dictate prompt and effective recall as opposed to any decision to avoid recall and assume the risk of exposure. It is rare that the engineers and quality control managers are held criminally liable. On the other hand, the chief executive officer and his people may be held liable and suffer under the law.

Specific references to sections of the law that affect individuals may be found in

1. Federal Trade Commission Act 15 USC S 41 *et seq.*, with specific reference to Section 45 (1)
2. U.S. Consumer Product Safety Act 15 USC S 2051 *et seq.*, with special reference to Sections 2068 through 2074
3. Magnuson-Moss Consumer Product Warranty Act 15 USC 2301 *et seq.*, with specific reference to Section 2310

The acts speak in terms of civil penalties of $10,000 and up for each *separate violation*. There are even stricter sections, larger fines, and criminal penalties in certain instances. The Food and Drug Act permits the sentencing of violators to time in prison for certain offenses and failure to conform to appropriate actions by the government.

Chapter XV

Warranty Interface

An important element of any recall is the maintenance of customer goodwill. To preserve this goodwill, the manufacturer must ensure that the customer's perception of the value of the product in question is, at a minimum, the same after the recall as before the recall. The customer must be left with the feeling that the recall was beneficial, or at least in no way detrimental, to his or her interests. One factor that will affect the customer's perception is how the recall affects the product's warranty versus how the customers think the recall should affect the warranty. This section of the guide will discuss some of the business decisions that will affect how the customer views the warranty-recall interaction.

Definition of a Warranty

Steven Gifis, in his *Law Dictionary*, defines warranty as an assurance by one party to a *contract* of the existence of a fact or facts upon which the other party may rely, intended precisely to relieve the promisee of any duty to ascertain the fact(s) for themselves, and which amount to a promise to indemnify the promisee for any loss if the facts warranted prove untrue.

The Uniform Commercial Code defines Express warranties.

Express warranties by the seller are created as follows:

1. Any affirmation of fact or promise by the seller to the buyer, which relates to the goods and becomes a part of the basis of the bargain, creates an express warranty—the goods shall conform to the affirmation or promise.
2. Any description of the goods that is made part of the basis of the bargain creates an express warranty that the whole of the goods shall conform to the description.
3. Any sample or model that is made part of the basis of the bargain creates an express warranty that the whole of the goods shall conform to the sample or model.

It is not necessary to the creation of an express warranty that the seller use formal words such as "warrant" or "guarantee" or that he have an intention to make a warranty, but an affirmation merely of the value of the goods or a statement purporting to be merely the seller's opinion or commendation of the goods does not create a warranty.

It is also said that unless excluded or modified, a warranty that the goods shall be merchantable is implied in a contract for sale, if the seller is a merchant with respect to goods of that kind.

A warranty is a promise on the part of the seller that the product will possess certain performance, safety, or quality characteristics.

This and the term "guarantee" are often confused and interchanged. What the combination of terms is often interpreted to mean is that the product or service will have certain characteristics and a definition of the manufacturer's, or guarantor's responsibility if the product fails to have one or more of the characteristics.

Types of Warranty

All warranties contain some type of promise of a level of performance, often couched in terms of "satisfaction," which can be expected to last for a specified minimum amount of time, that is, the one year warranty. The major difference between warranties lies in the "guaranty" section. These major variations in the "guaranty" portion are sufficiently significant to define the four following distinct types of warranty

- Parts and labor
- Parts only
- Labor only
- Percentage parts and labor

The "parts and labor" warranty states that for a specified period of time or cycles of operation after the purchase date, the warrantor will repair a defective product free of charge (depending on the economics of the individual case, the warrantor may elect to replace the item: this is also true for each of the types discussed below). Under this warranty the customer may be required to pay certain shipping and/or handling costs; these normally are the extent of the customer's liability.

The "parts only" warranty requires the warrantor to supply the parts only should a product be found defective or unsatisfactory. The shipping and handling charges may be the responsibility of the customer in addition to the labor charges incurred in the repair of the product. As with the a "parts and labor warranty," the parts warranty is in effect for a specified period of time. In some cases a "parts only" warranty may be attached to a parts and labor warranty. In this case the warrantor may bear the costs of parts and labor for (as an example) one year and the cost of required parts for an additional period of time, perhaps six months.

The "labor only" warranty is essentially the same as a "parts only" warranty except the labor costs are assumed by the warrantor and the parts costs are the responsibility of the customer.

The "labor only" warranty may also be attached to a parts and labor warranty as described above.

A "percentage" warranty can take the form either of the above three cases except that the warrantor is obligated to pay a specified percentage of the cost. A common example is the automotive starting battery which has a warranty that is, for example, five years long with the specification that after one year or some other short period the battery will be replaced at a fraction of the current listed cost depending upon the period the product has given service. If it has given 75 percent of the warranted service, 25 percent will be deducted from the replacement cost, and will be reduced a specified percentage each year up to the five year warranty time.

Is the Warranty in Effect?

In determining the interface between a product warranty and a recall of the product, the first question that must be asked is, "Has the warranty period expired?" The answer to this question will depend partly on the scope of the recall. Is only the product manufactured during the last several years being recalled, or does the recall cover many years including product recently manufactured? Is some of the recalled product under warranty? What effect will the warranty terms have on the recall, and what effect will the recall have on the warranty? These questions may be complicated legally. There may be liability for consequential damages, despite express disclaimers, years after the warranty has expired.

What Effect Will the Warranty Have on the Recall?

Once the decision to have a recall is made, the terms of the recall (who will pay for what?) must be established. If only product under warranty is involved, the shipping, handling and repair/replacements terms of the warranty may be adequate. If the reason for the recall is of such gravity, however, that immediate customer response is required, *other* techniques may be appropriate. For example, in an extreme case, it may be desirable to have a courier travel to the customer's site and pick up the product. In either event, many, if not all of the remedies and responsibilities described in the warranty can be invoked.

The most complex problem exists in the more general case where both products covered and not covered by the warranty are involved in the recall. In such a situation the warrantor may have four options as follows:

1. Utilize warranty terms for product under warranty or establish separate terms (probably less favorable to the other customers than the warranty terms).
2. Utilize the warranty terms for both warranted and non-warranted product.
3. Establish special terms (more favorable to the customers than warranty terms) for warranty product and less favorable terms for non-warranted product.
4. Establish special terms (as in 3) and apply to all products.

Which procedure is best will be determined by the answer to the following two questions.

"Will the customer's reaction to the recall terms be acceptable to the company?" A portion of the customers will always complain and be dissatisfied with the terms. Management must evaluate the alternatives and decide which ones will alienate the fewest customers, or perhaps more realistically, alienate only a small acceptable percentage.

The second question, closely related to the first, "What will the recall cost under these terms?" It must be verified that the terms are acceptable to the government agency demanding the recall. The question of cost versus customer satisfaction must be carefully considered, as it may determine whether the business is to survive. Is the company spending a minimum but alienating all of its customers, or is the cost of the recall so high that the recall costs are bound to have a serious affect on the company?

What Effect Will the Recall Have on the Warranty?

There is one final question. "What warranty will be given on the product to the customer?" If a warranty is issued, should it be

A continuation of the warranty in effect?

A new product warranty?

A shorter than new product warranty?

Or should there be no warranty at all?

If some of the product is under warranty and other customer's product is not under warranty, should a uniform policy be established or should a differential be established? What warranty will be issued to the older customers whose product has no warranty?

A useful factor can be included in the management's effort to answer the above questions is the nature of the product "fix." In other words, the action planned to correct the defect which is the reason for the recall can help management determine what warranty is issued on the returned product.

In a recall, the manufacturer has three corrective options: repair, replacement, or a refund. Obviously if the manufacturer opts for a refund, the warranty question does not arise. With the other two options, however, it will.

If the manufacturer chooses the repair option (which seems to be the most common), very few customers would expect a new product warranty. The customers whose product is under warranty will, however, expect some sort of a warranty. The question here is, for what period? If a repair warranty (if routinely given) extends further than the original, the customer will want the repair warranty; if not, they will want to continue the original warranty. It will require some sort of bookkeeping system entry to record who gets what warranty. If it not customary to issue repair warranties, the simplest system is merely to continue the original warranty, that is, status quo.

The customer whose product is not covered by a warranty may or may not be satisfied with the status quo. Deprived of the service of the product for a period of time, through no fault of their own, the customers may feel that they deserve some compensation for their inconvenience in the form of warranty. This feeling is particularly likely to occur when no repair warranty is issued.

In summary, the importance of customer goodwill versus the risk of customer dissatisfaction must be evaluated and steps taken to keep the customer goodwill at an acceptable level and balanced against cost.

Bibliography

American Insurance Service Group, Inc. "Controlling Product Liability Disasters," New York.

Bass, Lewis. 1994. *Managing for Products Liability Avoidance,* Chicago, IL: CCH Inc.

Consumer Product Safety Review, 1996. Special Investigations Unit Created to Aid Product Recall. CPSC 1: 1.

Corporate Legal Times. 1994. Roundtable Discussion: "Disaster Planning Can Minimize Crisis Effect." *Corporate Legal Times,* 39–49. Chicago, IL.

Food Safety and Inspection Service. 1988. U.S. Dept. of Agriculture Improving Recalls and the Food Safety and Inspection Service, Washington, DC.

Goodden, Randall. 1995. Preventing and Handling Product Liability. New York: Marcell Dekker.

Hutchinson, Joseph F., "Product Recall: Procedural Considerations." 1974 Proceedings, Product Liability Prevention Conference.

Jacobs, Richard M. and August B. Mundel. 1994. "Recall Planning." Proceedings of the 48th Annual Quality Congress, American Society for Quality Control. 826–833. Las Vegas, NV.

Moore, Michael Garth. 1995. "What is a Reasonable Safe Product? A Survey of Recalls and Retrofits." Park Ridge, IL: American Society of Safety Engineers.

Product Safety, Management Guidelines. 1989. National Safety Council, Chicago, IL.

Recall Handbook, A Guide for Manufacturers, Importers, Distributors and Retailers on Reporting under Section 15, 1988. U.S. Consumer Product Safety Commission, Washington, DC.

U.S.D.A. Chapter 7 Recalls and Emergency Procedures.

U.S.D.A. 1996. "Recall of Inspected Meat and Poultry Products." Decision to Recall, Washington, DC.

Index

A

B

C

R

S

T

U

V

W